Pre-Algebra Bingo Book

COMPLETE BINGO GAME IN A BOOK

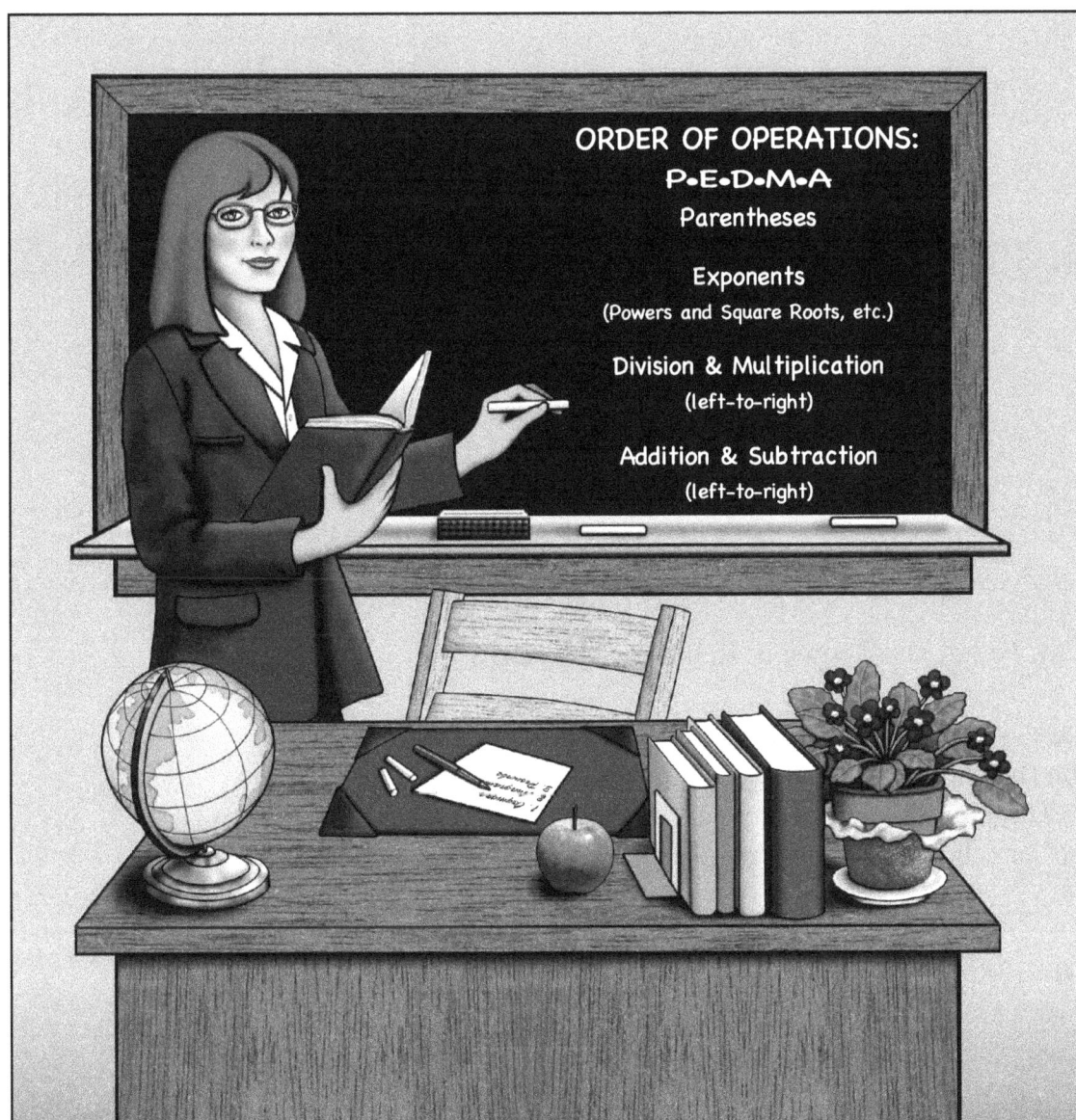

ORDER OF OPERATIONS:
P•E•D•M•A
Parentheses

Exponents
(Powers and Square Roots, etc.)

Division & Multiplication
(left-to-right)

Addition & Subtraction
(left-to-right)

Written By Rebecca Stark

Educational Books 'n' Bingo

TITLE: Pre-Algebra Bingo
AUTHOR: Rebecca Stark

ISBN 978-0-87386-459-6

Educational Books 'n' Bingo

Printed in the U.S.A.

PRE-ALGEBRA BINGO DIRECTIONS

INCLUDED:

List of Terms

Templates for Additional Terms and Clues

2 Clues per Term

30 Unique Bingo Cards

Markers

1. **Either cut apart the book or make copies of ALL the sheets. You might want to make an extra copy of the clue sheets to use for introduction and review. Keep the sheets in an envelope for easy reuse.**

2. Cut apart the call cards with terms and clues.

3. Pass out one bingo card per student. There are enough for a class of 30.

4. Pass out markers. You may cut apart the markers included in this book or use any other small items of your choice.

5. Decide whether or not you will require the entire card to be filled. Requiring the entire card to be filled provides a better review. However, if you have a short time to fill, you may prefer to have them do the just the border or some other format. Tell the class before you begin what is required.

6. There are 50 terms. Read the list before you begin. If there are any terms that have not been covered in class, you may want to read to the students the term and clues before you begin.

7. There is a blank space in the middle of each card. You can instruct the students to use it as a free space or you can write in answers to cover terms not included. Of course, in this case you would create your own clues. (Templates provided.)

8. Shuffle the cards and place them in a pile. Two or three clues are provided for each term. If you plan to play the game with the same group more than once, you might want to choose a different clue for each game. If not, you may choose to use more than one clue.

9. Be sure to keep the cards you have used for the present game in a separate pile. When a student calls, "Bingo," he or she will have to verify that the correct answers are on his or her card AND that the markers were placed in response to the proper questions. Pull out the cards that are on the student's card keeping them in the order they were used in the game. Read each clue as it was given and ask the student to identify the correct answer from his or her card.

10. If the student has the correct answers on the card AND has shown that they were marked in response to the *correct questions,* then that student is the winner and the game is over. If the student does not have the correct answers on the card OR he or she marked the answers in response to *the wrong questions,* then the game continues until there is a proper winner.

11. If you want to play again, reshuffle the cards and begin again.

Have fun!

TERMS/ANSWERS

-3	ALGEBRA
3/8	ALGEBRAIC EXPRESSION
1/2	ARITHMETIC
2	BINOMIAL
3	COEFFICIENT
4	COMMUTATIVE PROPERTY
5	CONSTANT
6	DISTRIBUTIVE PROPERTY
$6x - 30$	EQUATION
8	EQUIVALENT
9	EXPAND
10	EXPONENT
11	INVERSE OPERATION(S)
12	LINEAR EQUATION
$12xy$	MONOMIAL
14	NEGATIVE INTEGER
$15x$	PARENTHESES
18	POLYNOMIAL
21	POSITIVE INTEGER
$24 + n$	PROPORTION
$24x^2$	QUADRATIC EQUATION
$36x$	SIMPLEST FORM
$42x$	TERM(S)
55	TRINOMIAL
70	VARIABLE

Additional Terms

Choose as many additional terms as you would like and write them in the squares. Repeat each as desired.
Cut out the squares and randomly distribute them to the class.
Instruct the students to place their square on the center space of their card.

Clues for Additional Terms

Write three clues for each of your terms.

_____ 1. 2. 3.	_____ 1. 2. 3.
_____ 1. 2. 3.	_____ 1. 2. 3.
_____ 1. 2. 3.	_____ 1. 2. 3.

$2x + 5 = 9$	$2x + 5 = 9$	$2x + 5 = 9$	$2x + 5 = 9$	$2x + 5 = 9$
$2x + 5 = 9$	$2x + 5 = 9$	$2x + 5 = 9$	$2x + 5 = 9$	$2x + 5 = 9$
$2x + 5 = 9$	$2x + 5 = 9$	$2x + 5 = 9$	$2x + 5 = 9$	$2x + 5 = 9$
$2x + 5 = 9$	$2x + 5 = 9$	$2x + 5 = 9$	$2x + 5 = 9$	$2x + 5 = 9$
$2x + 5 = 9$	$2x + 5 = 9$	$2x + 5 = 9$	$2x + 5 = 9$	$2x + 5 = 9$
$2x + 5 = 9$	$2x + 5 = 9$	$2x + 5 = 9$	$2x + 5 = 9$	$2x + 5 = 9$
$2x + 5 = 9$	$2x + 5 = 9$	$2x + 5 = 9$	$2x + 5 = 9$	$2x + 5 = 9$

-3 What is the value of the variable? 1. $2(5x) + 15 = -15$ 2. $8x + 21 = -3$ 3. $24 \div x = -8$	**3/8** What is the value of the variable? 1. $1/8 = 1/3x$ 2. $2x = 3/4$ 3. $3y = 1\ 1/8$
1/2 1. If $y + 1/4 = 3/4$, then $y =$ ___. 2. If $x + 1/3 = 5/6$, then $x =$ ___. 3. If $1\ 1/2(x + 4) = 6\ 3/4$, then $x =$ ___.	**2** What is the value of the variable? 1. $5(x + 3) = 25$ 2. $3x + 5 = 2x + 7$ 3. $4(6y + 1) = 52$
3 1. If $4y = 12$, then $y =$ ___. 2. If $12x = 4$, then $x =$ ___. 3. If $y - 1/2 + 1/4 = 2\ 3/4$, then $y =$ ___.	**4** What is the value of the variable? 1. $6(11 - x) = 42$ 2. $1.5n = 6$ 3. $4/x = 1$
5 What is the value of the variable? 1. $-x - 4 = 4.5(-2)$ 2. $5x + 3x = 35 - x$ 3. $2y = 3y - 5$	**6** What is the value of the variable in each? 1. $x + 24 = 5x$ 2. Three times a number, n, is 12 more than n. 3. $7.5x = 45$
6x − 30 1. $6(x\text{-}5) =$ 2. $2x + 4x - 30 =$ 3. $2(3x - 15) =$	**8** What is the value of the variable in each? 1. $23x = 184$ 2. $2x + 3x = 48 - x$ 3. $x^2 + 4 = 68$

Pre-Algebra Bingo

9 What is the value of the variable in each? 1. $2(x + 3) = 24$ 2. $18/x = 2$ 3. $3y + 2(y + 1) = 47$	**10** What is the value of the variable in each? 1. $7x = 70$ 2. $1/2x = 5$ 3. $10y - 8y = 20$
11 What is the value of the variable in each? 1. $3x = 33$ 2. $x(3 + 2) = 55$ 3. $11/y = 1$	**12** What is the value of the variable in each? 1. $2/3 = 8x$ 2. $11/-3 = 44/-x$ 3. $12 + y = 4y/2$
12xy 1. $3x(4y) =$ 2 $4x(3y) =$ 3 $24xy \div 2 =$	**14** What is the value of the variable in each? 1. $x/7 = 2$ 2. $3x = 52$ 3. $4y + 12 = 68$
15x 1. $7x + 8x =$ 2. $3(5x) =$ 3. $45x / 3 =$	**18** What is the value of the variable in each? 1. $1/6x = 3$ 2. $.5x = 9$ 3. $3(y + 3) = 63$
21 What is the value of the variable in each? 1. $1/2(3 + x) = 12$ 2. $1/3x = 7$ 3. $3x + 2x = 105$ Pre-Algebra Bingo	**24 + n** The answer to these problems. 1. Mr. Jones was n years old 24 years ago. How old is Mr. Jones? 2. Ann is 24 years old. How old will she be in n years? 3. Jan has 24 books. Nan has n more than Jan. How many books does Nan have? © **Barbara M. Peller**

$24x^2$ 1. $8x(3x) =$ 2. $3x(8x) =$ 3. $12x(2x) =$	**$36x$** 1. $18x + 18x =$ 2. $45x - 9x =$ 3. $4(3y) + 6(4y) =$
$42x$ 1. $13x(4) =$ 2. $20x + 22x =$ 3. $84x \div 2 =$	**55** 1. In the equation $x - 35 = 20$. ___ is the value of x. 2. In the equation $x + 20 = 75$ ___ is the value of x. 3. If $x = 11$, then $5x =$ ___.
70 1. When a number, n, is divided by 14, the quotient is 5. What is that number? 2. The sum of 50 and a number, n, results in a sum of 120. What is that number? 3. When a number, n, is subtracted from 100, the difference is 30. What is that number?	**ALGEBRA** 1. One of the main branches of mathematics. 2. In this branch of mathematics, symbols, usually letters of the alphabet, represent unknown numbers. 3. This branch of mathematics works with variables
ALGEBRAIC EXPRESSION 1. A mathematical statement with at least 1 number, 1 variable, and 1 arithmetic operation. 2. $3x + 4$ is one; so is $4x -y$. 3. The algebraic expression to represent the number x added to 3 times the number y would be written as $x + 3y$.	**ARITHMETIC** 1. Its four main operations are addition, subtraction, multiplication and division. 2. Along with numbers and variables, an algebraic expression must include at least one ___ operation.
BINOMIAL 1. A polynomial with two unlike terms. 2. $5x - 3$ is one, but $5x^2$ is not. 3. $6y - 3$ is one, but $6x$ is not.	**COEFFICIENT** 1. The number to be multiplied by the variable. 2. In the term $4x$, the number 4 is the ___. 3. If a term consists of only a variable, such as the term xy, the ___ is 1.

COMMUTATIVE PROPERTY 1. This refers to the fact that the order in which two numbers are added or multiplied do not matter. 2. The fact that $5 + 2x = 2x + 5$ is an example of this property. 3. The fact that $5 \times 4 = 4 \times 5$ is an example of this property.	**CONSTANT** 1. A monomial term in an algebraic expression with no variables. 2. A term in an algebraic expression that contains only numbers. 3. In the algebraic expression $7 + 2x - y$ the number 7 is a ___.
DISTRIBUTIVE PROPERTY 1. This property lets you multiply a sum by multiplying each addend separately and then adding the products. 2. To use this property to multiply 5×42, multiply (5×40) and (5×2) and add the results. 3. An example of this property is $5 \times (3 + 4) = (5 \times 3) + (5 \times 4)$.	**EQUATION** 1. Any math sentence with an equal sign. 2. To solve one, perform the opposite operation on both sides of the equal sign. 3. An algebraic one includes at least one variable.
EQUIVALENT 1. We say two equations are this when they have the same solution. 2. We say fractions are this when they reduce to the same number. 3. 6/12 and 4/8 are ___ fractions; .5 and .50 are ___ decimals.	**EXPAND** 1. To distribute or multiply out the parts of an expression. 2. To ___ is the opposite of "to factor." 3. To carry out an indicated mathematical operation.
EXPONENT 1. The power to which a number or variable is raised. 2. In the term $7a^4$, the small number 4 is the ___. 3. In the term $8b^2$, the small number 2 is the ___.	**INVERSE OPERATION(S)** 1. An operation that undoes the effect of another. 2. Multiplication and division are ___. 3. Addition and subtraction are ___.
LINEAR EQUATION 1. In this kind of equation the greatest exponent of the variable is 1. 2. If a variable in an equation is raised to a power greater than 1, it is not this kind of equation. 3. $x + 3 = 7$ is one, but $x^2 + 3 = 7$ is not.	**MONOMIAL** 1. A polynomial with one term and no plus or minus sign. 2. $4xy^2$ is one, but $4xy^2 - 3$ is not. 3. $6x$ is one; so is $2y^2$. $2xy^2 + 5$ is not.

Pre-Algebra Bingo

NEGATIVE INTEGER
1. If you multiply a positive integer and a negative integer, the product will be a ___.
2. If you divide a negative integer by a positive integer, the quotient will be a ___.
3. If you divide a posiative integer by a negative integer, the product will be a ___.

PARENTHESES
1. Indicates terms that are one unit.
2. Symbols used in pairs to group things together.
3. First perform any calculations inside these symbols first.

POLYNOMIAL
1. Algebraic expression with constants and variables, including variables with whole-number, non-negative exponents.
2. This algebraic expression is a ___: $x^2 - 3x + 8$.
3. A ___ involves only the operations of addition, subtraction, multiplication, and non-negative integer exponents.

POSITIVE INTEGER
1. If you multiply two positive integers, the product will be a ___.
2. If you multiply two negative integers, the product will be a ___.
3. If you divide a negative integer by a negative integer, the product will be a ___.

PROPORTION
1. An equation with a ratio on either side of the equal sign.
2. It is a mathematical statement that says two ratios are equal.
3. If 1 of the 4 numbers in a ___ is a variable, use cross products to find the value of the variable.

QUADRATIC EQUATION
1. An equation in which one or more variables in an equation is raised to the power of 2 but no higher.
2. $2x^2 + 3x = 14$ is this type of equation, but $x^3 + 3x = 14$ is not.
3. $x^2 + y^2 = 13$ is this type of equation, but $x^3 + y^3 = 35$ is not.

SIMPLEST FORM
1. A algebraic expression is in its ___ when it contains no parentheses or like terms.
2. The ___ of the expression $2x + 3x$ is $6x$.
3. The ___ of the expression $2x + 4x + 2y + 3y$ is $6x + 5y$.

TERM(S)
1. A single number or variable or a number and variable multiplied together.
2. In the expression 4x – 8, 4x and 8 are these.
3. In the equation $3x + 2x + y = 13$, 3x and 2x are called like ___ because they differ only by a constant factor.

TRINOMIAL
1. A polynomial with three unlike terms and two plus and/or minus signs.
2. $5x - 3 + 5y$ is one, but $10x^2$ is not.
3. $6y + 3 + 2x$ is one, but $6x - 3$ is not.

VARIABLE
1. A letter that stands for an unspecifed value to be replaced.
2. In the equation $x - 2 = 3$ the letter x is one.
3. In the equation $12y - 4y = 64$ the letter y is one.

Pre-Algebra Bingo

Linear Equation	3/8	3	55	Distributive Property
Commutative Property	2	Equivalent	11	Algebraic Expression
1/2	Simplest Form		Binomial	Inverse Operation(s)
4	12*xy*	Term(s)	14	Parentheses
Positive Integer	21	42*x*	Polynomial	Monomial

Pre-Algebra Bingo: Card No. 1

Pre-Algebra Bingo

4	1/2	Exponent	Coefficient	12
Parentheses	11	6	12*xy*	Arithmetic
8	21		15*x*	Term(s)
Equation	Expand	Simplest Form	Quadratic Equation	Monomial
Algebraic Expression	Equivalent	42*x*	Commutative Property	Polynomial

Pre-Algebra Bingo

4	Term(s)	11	14	1/2
21	2	6x - 30	3/8	Factors
12xy	Equivalent		Arithmetic	-3
Simplest Form	8	Positive Integer	Equation	Exponent
Polynomial	Commutative Property	42x	Quadratic Equation	12

Pre-Algebra Bingo

Simplest Form	Arithmetic	3	Commutative Property	Linear Equation
Algebra	5	3/8	Coefficient	1/2
Binomial	Equation		Distributive Property	55
Term(s)	18	Equivalent	$42x$	6
$24x^2$	Algebraic Expression	Trinomial	Polynomial	Inverse Operation(s)

Pre-Algebra Bingo

Algebraic Expression	Distributive Property	12xy	6	Commutative Property
Algebra	Term(s)	6x - 30	15x	2
3	Inverse Operation(s)		Negative Integer	24 + n
Monomial	12	Linear Equation	Quadratic Equation	10
11	42x	1/2	Simplest Form	Binomial

Pre-Algebra Bingo: Card No. 5

Pre-Algebra Bingo

-3	Arithmetic	Exponent	12	Inverse Operation(s)
14	12xy	10	3/8	1/2
Coefficient	24x^2		5	15x
42x	Positive Integer	Quadratic Equation	Trinomial	3
Parentheses	Term(s)	Linear Equation	Binomial	18

Pre-Algebra Bingo

Linear Equation	Arithmetic	$24 + n$	Negative Integer	11
Parentheses	12	21	2	Algebra
Exponent	55		$15x$	5
Simplest Form	Equation	$6x - 30$	4	8
$42x$	Commutative Property	Quadratic Equation	Trinomial	-3

Pre-Algebra Bingo

Binomial	Arithmetic	9	14	5
Algebra	3	Coefficient	Inverse Operation(s)	6
18	Proportion		12	Distributive Property
Polynomial	Simplest Form	4	$24x^2$	Equation
Equivalent	$42x$	Trinomial	$12xy$	Parentheses

Pre-Algebra Bingo

$15x$	11	21	18	Commutative Property
$24x^2$	12	Binomial	$12xy$	Arithmetic
Factors	Linear Equation		2	9
10	Monomial	Positive Integer	Negative Integer	$24 + n$
Equation	Quadratic Equation	$6x - 30$	4	Distributive Property

Pre-Algebra Bingo

4	14	5	Coefficient	18
Inverse Operation(s)	6	3/8	2	12
Proportion	Arithmetic		55	8
Positive Integer	Monomial	10	Quadratic Equation	Factors
6x - 30	Parentheses	Exponent	Algebraic Expression	Binomial

Pre-Algebra Bingo

-3	Arithmetic	12xy	10	Parentheses
9	Factors	Negative Integer	15x	3/8
Algebra	12		Exponent	21
6x - 30	1/2	Quadratic Equation	Commutative Property	4
24x^2	42x	Linear Equation	Trinomial	11

Pre-Algebra Bingo: Card No. 11

Pre-Algebra Bingo

11	Distributive Property	Factors	14	15x
21	Parentheses	3	Trinomial	2
Linear Equation	24 + n		Inverse Operation(s)	Coefficient
42x	Equation	12	4	Algebra
Arithmetic	9	Proportion	24x^2	6

Pre-Algebra Bingo

10	Distributive Property	-3	Factors	Inverse Operation(s)
3	9	12	15x	8
14	6		21	24 + n
Binomial	Quadratic Equation	5	Proportion	4
42x	Monomial	Trinomial	Linear Equation	Negative Integer

Pre-Algebra Bingo

Commutative Property	12	12xy	15x	24x^2
6	Linear Equation	Factors	2	Arithmetic
10	55		Exponent	6x - 30
Monomial	Quadratic Equation	Proportion	5	-3
42x	Coefficient	8	Parentheses	Binomial

Pre-Algebra Bingo

Negative Integer	$15x$	$12xy$	11	14
-3	Exponent	3/8	3	$24x^2$
Inverse Operation(s)	Linear Equation		1/2	Arithmetic
$42x$	Factors	9	Quadratic Equation	10
Parentheses	Equation	Trinomial	18	21

Pre-Algebra Bingo

5	Factors	9	18	Expand
Coefficient	8	$24 + n$	Algebra	55
10	Distributive Property		Inverse Operation(s)	21
Simplest Form	6	$42x$	Negative Integer	4
$24x^2$	Constant	Trinomial	Equation	Arithmetic

Pre-Algebra Bingo

$6x - 30$	Variable	$36x$	Factors	Commutative Property
Negative Integer	$24x^2$	Quadratic Equation	55	$24 + n$
$15x$	Binomial		Constant	9
Monomial	Parentheses	4	$12xy$	8
Positive Integer	10	11	14	Distributive Property

Pre-Algebra Bingo

18	Proportion	6	10	Coefficient
Arithmetic	6x - 30	Positive Integer	Inverse Operation(s)	24x^2
15x	8		36x	3
Monomial	3/8	Quadratic Equation	1	Exponent
Constant	Factors	12xy	Variable	-3

Pre-Algebra Bingo

Inverse Operation(s)	-3	Factors	9	Proportion
Negative Integer	14	Arithmetic	11	55
Variable	Commutative Property		2	1/2
Exponent	Constant	Positive Integer	Equation	36x
3	Expand	Parentheses	Binomial	Trinomial

Pre-Algebra Bingo

Proportion	Variable	14	Factors	Trinomial
6	21	Algebra	Positive Integer	Coefficient
Distributive Property	24 + n		Simplest Form	3/8
Algebraic Expression	Equivalent	Polynomial	Equation	Constant
Term(s)	Binomial	Expand	4	36x

Pre-Algebra Bingo

Negative Integer	-3	Algebra	Factors	Algebraic Expression
Distributive Property	36x	5	9	24 + n
8	Parentheses		Variable	12xy
Positive Integer	11	Constant	Monomial	Binomial
Simplest Form	Expand	Trinomial	6x - 30	Equation

Pre-Algebra Bingo

18	Exponent	36x	3	10
Coefficient	14	1/2	9	2
6	55		Linear Equation	24 + n
Constant	Monomial	Equation	3/8	Commutative Property
Expand	6x - 30	Variable	8	Algebra

Pre-Algebra Bingo

5	Variable	11	3	Trinomial
-3	Proportion	Parentheses	Negative Integer	3/8
Exponent	10		Polynomial	Linear Equation
8	Expand	Constant	6x - 30	Equation
Algebraic Expression	Equivalent	Binomial	Positive Integer	36x

Pre-Algebra Bingo

5	Proportion	Commutative Property	Variable	9
36x	Trinomial	Algebra	Coefficient	Linear Equation
24 + n	18		10	8
Algebraic Expression	Polynomial	Constant	6x - 30	Distributive Property
Term(s)	Simplest Form	Expand	14	Equivalent

Pre-Algebra Bingo

Simplest Form	Algebra	Variable	12xy	36x
3/8	Monomial	Negative Integer	5	2
Distributive Property	9		Polynomial	Constant
1/2	Algebraic Expression	Equivalent	Expand	55
Trinomial	Commutative Property	6	24x^2	Term(s)

Pre-Algebra Bingo

36x	Variable	Exponent	Coefficient	18
Positive Integer	14	9	Proportion	5
Monomial	Polynomial		55	Simplest Form
6x - 30	3	Algebraic Expression	Expand	Constant
24 + n	24x^2	12xy	Equivalent	Term(s)

Pre-Algebra Bingo

Exponent	6x - 30	Variable	Proportion	21
Algebraic Expression	Polynomial	Negative Integer	Constant	2
Quadratic Equation	Equivalent		Expand	Simplest Form
18	-3	Algebra	Term(s)	3/8
24x^2	55	36x	1/2	24 + n

Pre-Algebra Bingo

Inverse Operation(s)	Proportion	1/2	Variable	5
21	36x	Polynomial	Coefficient	55
Equivalent	8		24 + n	Positive Integer
4	18	Parentheses	Expand	Conctant
3	15x	24x^2	Term(s)	Algebraic Expression

Pre-Algebra Bingo

36x	Proportion	18	Negative Integer	15x
Monomial	Positive Integer	Algebra	24 + n	1/2
Distributive Property	Polynomial		2	Variable
21	Algebraic Expression	12	Expand	Constant
5	9	Term(s)	-3	Equivalent

Pre-Algebra Bingo: Card No. 29

Pre-Algebra Bingo

Commutative Property	Variable	Coefficient	15x	Constant
3/8	Proportion	Exponent	55	2
Monomial	10		24 + n	Algebra
Term(s)	-3	3	Expand	Polynomial
Algebraic Expression	11	Equivalent	36x	1/2